Fayera Debel

The impact of population growth on forestry development in east Wollega zone

The case of haro limu woreda

GRIN Verlag

Bibliografische Information der Deutschen Nationalbibliothek:

Die Deutsche Bibliothek verzeichnet diese Publikation in der Deutschen Nationalbibliografie; detaillierte bibliografische Daten sind im Internet über http://dnb.d-nb.de/ abrufbar.

Imprint:

Druck und Bindung: Books on Demand GmbH, Norderstedt Germany
ISBN: 978-3-656-71826-0

This book at GRIN:

http://www.grin.com/en/e-book/278848/the-impact-of-population-growth-on-forestry-development-in-east-wollega

THE IMPACT OF POPULATION GROWTH ON FORESTRY DEVELOPMENT IN EAST WOLLEGA ZONE, THE CASE OF HARO LIMU WOREDA

BY

FAYERA DEBEL GASHE

August, 2014

ACKNOWLEGDEMENT

This study would never be completed without the contribution of many people to whom I would like to express my thanks. Many thanks and appreciation goes to individuals whom without their help and support, the successful completion of my study would not have been possible. I am very grateful to my major supervisor, Dereje Chimdesa for his guidance and encouragement to accomplish this work. I would like to praise my God for his unspeakable gift in all my life in general and his help in the process of undertaking this study in particular.

TABLE OF CONTENTS

LIST OF TABLES

Page

LIST OF GRAPHS

ABREVIATIONS AND ACRONYMS

CSA	Central Statistical Agency
FAO	Food and Agricultural Organization
GDP	Gross Domestic Product
HCVF	High Conservation Value Forest
HLDFEDO	Haro Limu Finance and Economic Development Office
US	United States

ABSTRACT

Deforestation is a growing problem in many parts of the tropical world and one of the affected countries is Ethiopia. The general objective of this study is to assess the effect of population growth on forest resource in East Wollega Zone in general and Haro Limu woreda in particular. The data used for the study were collected from 89 farm households heads drawn from the four kebeles of Haro Limmu district. Probability proportional to size sampling technique was employed to select the farm households from four peasant associations, which were selected by random sampling techniques. Primary data were collected using a structured questionnaire. In addition, secondary data were extracted from relevant sources to supplement the data obtained from the survey.

The result of this study reveals that population growth huge impact on forestry development in the ways of expanding agricultural land, using wood as energy sources and satisfying the input requirements in agricultural activity. Respondents use family planning services in reducing the impact of population growth on the forestry development

CHAPTER ONE

1. INTRODUCTION

1.1. Background of the Study

Ethiopia with its 79 million (CSA, 2008) people living in a geographical extent of 1.1 million km2 has a GDP of US $ 6.1 billion, 39% of which is contributed by Agriculture, upon which 85% of the population are dependent for livelihoods (Amare, 2013).

Forest is the resource that provides many benefits to the society. In addition to providing wood, forest provide a habitat for wild life, site for recreation, wildness, watershed protection and many other benefits. It can also absorb atmospheric carbon dioxide (Dealon, 1993).

The impact of demographic changes on forests and the environment is often discussed in terms of biological carrying capacity, i.e. the maximum number of individuals that a resource can sustain. However, many factors influence carrying capacity, such as economic development, socio-political processes, and trade, technology, and consumption preferences (Bijendra, 2009).

There has been a steady increase in the population growth rate since 1960 but after 1960's population in Ethiopia started to increase rapidly because of changes in socio-economic conditions in the world (Amare, 2013).

According to Amare (2013) parallel to low socio-economic and rapid population growth, there is land degradation which has affected the country's ecological integrity. For instance, the forest cover of Ethiopia was 40% at the beginning of 20th century but has declined to 2.2% at present. Soil degradation process makes large areas unsuitable for agriculture. Because the top soil and even part of the sub soil in some areas has been removed, and stones or bare rock are left at the surface. Up to half of the arable land in the Ethiopian highlands is estimated to be moderately to severely eroded, and as a result, previously cultivable lands are being (or have been) turned to wasteland (Amare, 2013).

Rapid population growth and the low economic standard of living in Ethiopia have brought in their awake numerous consequences to land cover and use changes, change in climate and hydrological status in the country. In Ethiopia studies have indicated that as there is agricultural land expansion at the expense of other land uses. Ethiopian highlands are fragile and forest, water and its biodiversity is climate dependant, which are currently under stress due to population pressure and mismanagement of natural resources. The poverty stricken economy of Ethiopia needed scientific management of its natural resources and balancing the population to cope up with the climate change and the challenges of the globalization of economy. Thus, there is a need to review the population and environment interrelation at the country level in order to suggest the means to minimize adverse effect of population pressure on environment (Amare, 2013).

The general objective of this study is to assess the effect of population growth on forest resource in East Wollega Zone in general and Haro Limu woreda in particular.

1.2. Statement of the Problem

Deforestation is a growing problem in many parts of the tropical world and one of the affected countries is Ethiopia. According to World Bank (2009), Ethiopia which has a population of about 81 million is the second most populous country in Sub-Saharan Africa. Ethiopia is also one of the world's poorest countries. In addition, the spread of poverty, the environmental degradation especially deforestation is due to local clearing of forest for their personal needs, such as for fuel, hunting, agriculture, and for religious reason. Deforestation is indiscriminate cutting or over-harvesting of trees for lumber or pulp, or to clear the land for agriculture, ranching, construction, or other human activities (Encarta, 2009)

The direct causes of land degradation are mainly deforestation, overgrazing and over-cutting, shifting cultivation and agricultural mismanagement of soil and water resources: such as non-adoption of soil and water conservation practices, improper crop rotation, use of marginal land, insufficient and/or excessive use of fertilizers, mismanagement of irrigation schemes and over pumping of groundwater. The indirect causes of land degradation are mainly population increase, land shortage, short-term or insecure land tenure and poverty and economic pressure (FAO, 2001). The economic contribution of forest are associated with the production, consumption of goods and services , the supply of fuel wood, construction materials and timber come mainly from forest (Negash, 2003).

The rapid deforestation or the declining of forest accompanied by variety of other environmental problems is caused by agricultural expansion and wood gathering. The major factors contributing to deforestation are poverty, low level of income and population growth.

For this reason households have to depend on clouding for their fuel consumption and this is hampering the farmers from using organic manure as fertilizer for their already degraded farmlands. The wide range of topography has contributed to the presence of enormous biological and cultural diversity. One of the solutions proposed to arrest forest degradation in certain area is to introduce the practices of participatory forest management by involving the local people. Therefore, this study attempts to show the effect of high population growth on the natural forest in the Haro Limu and also tries to give some analysis based on finding.

1.3. Objective of the Study

1.3.1. General Objective

The general objective of this study is to assess the effect of population growth on forest resource in East Wollega Zone in general and Haro Limu woreda in particular.

1.3.2. Specific Objectives

Specific objectives of this study are to:

- Review the overall importance of forest in relation to growing population
- Examine the linkage between population and deforestation
- Identify the major causes and consequences of deforestation

1.4. Hypothesis

One can see from theoretical and practical evidence that any programs of development approach with inevitable by encountered a number of obstacles. By the same taken impact of population growth on forestry in particular can be limited by many factors.

In carrying out this study, it can be hypothesized that:

- Most of the population growth members participating in forestry which may be affect forest or not.
- There is impact of population growth on forestry in this area to decided the effect of the forest beneficiary.

1.5. Significance of the Study

The environmental benefits of forest are immense. The forest provides the economy with raw materials which are transformed into consumer products by the production process. Ultimately these raw materials and energy return to the environment as a waste product. Forest resource also plays essential role in nutrient cycling, carbon cycling and hydrological cycling.

However the high rate of deforestation in woreda to reduce all these important ecological benefits of forest. Deforestation also affects important economic activities such as agriculture. Therefore, it is important to study the cause, consequences of deforestation in woreda.

1.6. Scope of the Study

Population growth has a number of influences on a country i.e. economically, socially, psychologically, politically and etc. But this study was limited to show the impact of population growth on the ecosystem.

1.7. Limitations of the Study

While conducting this study the following limitation as faced;

- Shortage of time to gather the data relevant to the study.
- There is shortage of money in data collection

Although the study was limited both in sample size, time and area coverage, the results of the study was expected to be of value in examining the effect of population growth on forest resource.

CHAPTER TWO

2. REVIEW OF LITERATURE

2.1. Theoretical Literature

2.1.1. Meaning and Value of Forest

Forestry is management of forestlands for maximum sustained yield of forest resources and benefits. Although forestry originally concerned mainly timber production, it now also involves the management of grazing areas for domestic livestock, the preservation of wildlife habitats, watershed protection, and the development of recreational opportunities. The management of forestlands therefore helps to ensure that wooded areas are used for maximum benefit according to their nature (Encarta, 2009).

Forests worldwide are known to be critically important habitats in terms of the biological diversity they contain and in terms of the ecological functions they serve (Hamdallah, 2001). The ecological services of forests are similarly many. Forests regulate local and global climate, ameliorate weather events, regulate the hydrological cycle, protect watersheds and their vegetation, water flows and soils, and provide a vast store of genetic information much of which has yet to be uncovered. Scientists debate the linkages between biological diversity and ecological services. Those who believe in a strong link argue that any ecosystem, forests included, cannot cope with stresses and shocks if the diversity of the system has been reduced. Others argue that a majority of species are 'redundant' in the sense that their removal would not impair ecosystem functioning (Hamdallah, 2001).

Every forest has some environmental and social value. The values it contains may include rare species, recreational sites or resources harvested by local residents. Where these values are considered to be of outstanding significance or critical importance, the forest can be defined as a High Conservation Value Forest (HCVF) (Steve *et al.*, 2003).

2.1.2. Causes of Deforestation

Deforestation is the conversion of forest to an alternative permanent non-forested land use such as agriculture, grazing or urban development (van Kooten and Bulte, 2000). Deforestation is primarily a concern for the developing countries of the tropics (Myers, 1994) as it is shrinking areas of the tropical forests (Barraclough and Ghimire, 2000) causing loss of biodiversity and enhancing the greenhouse effect (Angelsen *et al.*, 1999). FAO considers a plantation of trees established primarily for timber production to be forest and therefore does not classify natural forest conversion to plantation as deforestation (but still records it as a loss of natural forests). However, FAO does not consider tree plantations that provide non-timber products to be forest although they do classify rubber plantations as forest. Forest degradation occurs when the ecosystem functions of the forest are degraded but where the area remains forested rather cleared.

The struggle to save the world's rainforests and other forests continues and there is a growing worldwide concern about the issue. In order to save forests, we need to know why they are being destroyed. Distinguishing between the agents of deforestation and its causes is very important in order to understand the major determinants of deforestation. The agents of deforestation are those slash and burn farmers, commercial farmers, ranchers, loggers, firewood collectors, infrastructure developers and others who are cutting down the forests. Causes of deforestation are the forces that motivate the agents to clear the forests (Chakravarty *et al.*, 2012).

However, most of the existing literature typically distinguishes between two levels of specific factors: direct and indirect causes of deforestation. Direct agents and causes of deforestation, also typically referred to as sources of deforestation, first level or proximate causes are relatively easy to identify but the indirect causes which are usually the main divers of deforestation are the ones that cause most disagreement and the ones that are hardest to quantify (Chakravarty *et al.*, 2012).

2.1.3. Connections between Population and the Environment

With Africa and Asia expected to account for the bulk of global population growth during the coming decades, the international community has become concerned over the looming scarcity (and/or degradation) of the natural resources needed to support people in these continents, particularly in sub-Saharan Africa, South Asia, and East Asia. The future availability of resources — whether soil, woodlands, or water—are expressed in per capita terms (Geoffrey, 2011).

Many countries whose per capita food consumption falls shorts of basic nutritional requirements will continue to experience significant population growth in the coming decade, thereby increasing the number of people without access to sufficient food.

According to Geoffrey (2011), as populations grow across the developing world, so too will carbon emissions assuming other factors do not change. However, emissions levels in most developing nations pale in comparison to emissions levels in industrialized countries.

While per capita approaches provide insight into the connections between population growth and environmental security, they present an incomplete picture of the ways demographic change affects both local ecosystems and global environmental trends. Focusing on per capita data often obscures variable consumption levels associated with wealth and fails to distinguish between relevant demographic characteristics, such as different resource use patterns between urban and rural populations. To gain a more nuanced understanding of the connection between resource depletion and population growth, it is necessary to recognize environmental resources as fundamentally interlinked. Viewing environmental conditions through a holistic lens is the only way to develop an accurate picture of the situation on the ground, either globally or in a particular region (Geoffrey, 2011).

As populations increase, so does food demand, which imposes pressure on soil, water, and woodland resources. For instance, the need for heightened agricultural productivity burdens surface and groundwater resources because of irrigation needs. Then, the drive for increased crop yield often results in intensive and unsustainable use of farmland, inflicting long-term damage to soil quality.

Also, rising food demand regularly necessitates clearing new land for agriculture, which commonly means felling forests, thus accelerating soil erosion and interrupting the hydrological cycle that spurs rainfall (Geoffrey, 2011).

Deforestation caused by increased resource consumption hinders woodland areas' ability to serve as "carbon sinks" that pull carbon from, and release oxygen into, the atmosphere. This hinders forests' ability to act as natural mitigators against greenhouse gas emissions. Because of this, decline in global forest cover both accelerates and deepens the ramifications of climate change.

Viewing the interconnectedness of resource degradation through a food-security lens is instructive, but it does not account for all factors. Economic modernization, rising living standards, and changing consumer preferences across the developing world also place a substantial burden on natural resources (Geoffrey, 2011).

2.2. Empirical Literature

2.2.1. Trend in Population Size and Growth

In developing countries, land degradation is one of the greatest threats which strikes at the basic resource of the population. The degradation process is so acute in Africa that millions of people have fallen into poverty and have suffered famine and death. The situation in Ethiopia is not different. In Ethiopia, massive environmental degradation has occurred during the last few decades due to natural factors, unwise use of its natural resources, unsound ecological practices and population pressure. The major factor accelerating the environmental degradation rate is man's abusive actions such as the removal of the natural vegetation cover through deforestation, over-grazing and inappropriate agricultural practices. The population which is growing at a very rapid rate of about 3 percent annually has been clearing forests and vegetation at an alarming rate in order to meet its increasing requirements of food, fiber and energy (Carla et al., 2001).

counted population of Haro Limmu district based on population and housing census conducted in 2012 G.C is 61,235 and 62,750 in 2003 E.C and 2004 E.C respectively. By the year 2004 E.C from 62,750 total populations of the district 31,340 (9.58%) were males where as about 31,410 (9.63%) were females. During this year about 10.25% of the total populations were rural population, which are directly engaged their life with even the back bone of the country called agriculture. The crude population density of the district in the year 2004 E.C was **23.1** persons per. km^2 (HLDFEDO, 2011).

2.2.2. Pattern of Deforestation in Haro Limu

Climate, the long-term effect of the sun's radiation on the rotating earth's varied surface and atmosphere. It can be understood most easily in terms of annual or seasonal averages of temperature and precipitation. Most part of the land has an elevation above 1400 meters and characterized. By sub tropical climatic condition with a mean annual temperature between 14.90 c and 34.90 and mean annual rainfall of 1200mm to 2600mm (HLDFEDO, 2011).

Major type of natural vegetation includes high forest covering 4,372 hectares exist in the district. The other type of natural vegetation found in the district is shrub and bush land with 1,222.75hectare and manmade forest on 832.26 hectares of land also exists in the district (HLDFEDO, 2011).

Farmers of the district used the two methods of soil fertility. Traditional methods of maintaining soil fertility used are addition of manure, fallowing, crop rotation and intercropping where as modern methods of maintaining soil fertility in the district are using adding chemical fertilizers and use compost. Planting and check dams are among traditional methods of soil conservation and soil band, cut off drain, mixed farming, allay cropping and water way are modern methods of soil conservation exist in the district (HLDFEDO, 2011).

2.2.3. Economic and Ecological Significance of Forest Resource

The term land use refers to the ways that people use land and the natural resources it provides. It is the best allocation of land for its best alternative uses. Land use potential is necessary to select the land characteristics needed for any production. Some of the major factors that determine the potentiality of the land are temperature, length of growing period, moisture availability, flood hazard, degradation hazard, toxicity, rooting condition and workability.

Out of the total land of the district the proximate areal coverage of land used for crop cultivation is 45,659 hectares of which 40,885 hectares of land is used for annual crop cultivation and 863 hectares of land is used for perennial crop production.

Arable land is a land that is ideal and economical for the cultivation of crops. Arable land is an area with more than 90 days of dependable growing period, soil depth of more than 25 cm and surface stoniness of less than 50 to 90 %. Arable is pertaining to tillable land that is suitable for tillage and crop production. The area of arable land used in the district is estimated to be 40,714.25 hectares of land (HLDFEDO, 2011).

The Natural forest of the district covers the total area of 8,312.11 hectares of land. Manmade type of forest is planted to solve the problem of environmental problem such as soil erosion, desertification, deforestation, and etc. With the aim of satisfying one of the millennium development goals of United Nations the inhabitants of the district were participated on the planting and protecting the trees. Out of the total land of the district about 1,275 hectare is covered with manmade forest. The bush land of the district covers the total area of 188,555 hectare of land (HLDFEDO, 2011).

CHAPTER THREE

3. RESEARCH METHODOLOGY

3.1. Description of the Study Area

Haro Limmu is one of the districts of East Wollega zone. Today this district is sub divided in to 15 farmers associations and one urban center for all its administrative purposes.
Before the year 1998, this district was under the district Limmu but later on Haro Limmu became an independent district with the administrative center at Hero. Being one of the jawi's children, Haro is living around Anger river. Jaws belong to Oromo clan called mocha. Haro is also a woman who where living at a place now a day called Haro town. In Haro youth is also strong & severe, this is because of low soico- economic status of family to manage the youth, limited investment, lack of meaning full education and training, inability to incant work ethic in the youth cultural and psychological constitution, non expansion of investment and mass media infrastructure. The name of this district was derived from the combination of the jawi family Haro and Limmu. Finally, Haro Limmu district was established December 21/4/1998 E.C (HLDFEDO, 2011).

This district is contiguous with Limmu in the east, Benshangul Gumuz Regional state in west and Anger River in the south and Ebantu district in the north of the district. This district was located at distance 165 km & 488 km from zonal town called nekemte and Addis Ababa respectively. Today this district is sub divided into 15 farmers associations 1 urban kebele with a capital town of Haro for all it administrative purposes.

In the beginning of the 20th the district was known by the name Haro Limmu but later on Ebantu became an independent district & then Haro Limmu was established as independent district with the administrative center at Haro and Limmu district with the administrative center at Gelila respectively (HLDFEDO, 2011).

3.2. Demographic Characteristics of the Study Area

The majority of the population of the district were included in the age group 15-64 and family size of the district were 32,739 for rural and 3,760 for urban. Based on the population density there is dispersed rural settlement pattern in each peasant association. The targeted population of the study was all of in this district.

Table 3.1 population of the district by age group

Age group	Rural			Urban			Total		
	Male	Female	Total	Male	Female	Total	Male	Female	Total
0-14	11,598	11,452	23,050	1,052	953	2 005	12,650	12,405	25,055
15-64	16,448	16,291	32,739	1,955	1 805	3,760	18,403	18,096	36,499
>= 65	950	806	1,756	150	102	252	1,100	908	2008
Total	8,996	28,549	57,545	3,157	2 ,860	6,017	32,153	31,409	63,562

Source: Haro Limmu District Agricultural Development Office, 2005 E.C

3.3. Data Sources of data and Methods of Data Collection

3.3.1. Type and Source of Data

In order to analyze the stated objectives both primary and secondary data was used. The main source of data for the study was sample household survey data to be collected from randomly selected households in the study districts. The data was expected to consist of sample household survey, household attributes and socio-economic characteristics and other related information in the study area that is essential for this study. The secondary data sources are annual reports and plans of zonal and the respective district offices.

3.3.2. Methods of Data Collection

Structured questionnaire was used to collect primary data. Detailed information on household composition and characteristics, environmental problems, and institutional factors was collected through interview of sample household heads.

3.4. Sampling Technique and Sample size

3.4.1. Sampling Technique

The study was conducted in Haro Limu district, East Wollega Zone of Oromia Regional State. This district was purposefully selected due to the fact that in this area the environment has been degraded largely because of settlement made from other area. Random sampling technique was employed to draw sample of household heads. From a total of 15 peasant associations in this district four peasant associations were selected randomly.

3.4.2. Sample Size

After deciding the sample area and sampling design it is important to fix the sample size, as it contributes to the cost and reliability of the study. From 15 peasant associations in the district, 4 peasant associations were sampled using simple random sampling procedure which accounts for almost 27 % of the total peasant associations of the study area. Depending on the availability of time and budget to do this paper I was used about 89 households and this was proportionally distributed for selected kebeles based on their number of households.

Table 3.2 Number of households and sample size

Peasant Association	Number of household heads	Sample size
Sombo Gadisa	2,286	16
Homi Kelala	5,387	37
Gorba Gudina	4,033	28
Rifenti	1,128	8
Total	**12,834**	**89**

For systematic random sampling of respondents the sample interval for picking each respondent the following formula was used. $k = \frac{N}{n} = \frac{12{,}834}{89} = 144$ Where n is sample size, N is number of population and k is the skip interval between sampled farmers (Kultar, 2007). Each sample was taken at the interval of 144 and the 1st will be drawn using lottery method

3.5. Data Processing and Analysis

The relevant data being collected was processed and analyzed with tables and percentages and the data analysis method used was descriptive statistics because of its simplicity and clarity.

During analysis and interpretation the following was considered.

- Reasonable explanation for the relations was made.
- Interpreting tasks was accomplished only after considering relevant factors affecting the problem

CHAPTER 4

4. RESULT AND DISCUSSION

This chapter deals with the analyses and interpretation of major findings of the study on the impact of population growth on forestry development in Haro Limu districts. From all of the respondents 17.98 % were from Sombo Gadisa kebele, 31.46 % were from Gorba Gudina, 41.57 % were from Homi Kelala and 8.99 % were from Rifenti.

4.1. Impact of population growth on forest

Importance of forests varies in the daily life of people living close to it. For instance, rural populations depend most fundamentally on forests in terms of subsistence, health, income and culture.

Table 4.1. Participation in controlling impact of population growth on forest

		Participation in controlling impact of population growth on forest		Total
		Yes	No	
Sex of respondents	Female	9	0	9
	Male	67	13	80
	Total	76	13	89

Source: Own computation

From the above table it can be observed that about 76 respondents have been participated in controlling the impact of population growth on forest while the rest of respondent have no care about forest development in the district under study.

Table 4.2 Participation in controlling impact of population growth on forest

		Participation in controlling impact of population growth on forest		Total
		Yes	No	
Educational level	Illiterate	5 (83%)	1 (17%)	6
	<=Grade 4	31 (91%)	3 (9%)	34
	Grade 5-8	17 (85%)	3 (15%)	20
	Grade 9-10	9 (69%)	4 (31%)	13
	Grade 11-12	8 (80%)	2 (20%)	10
	Diploma and above	6 (100%)	0	6
Total		76	13	89

Source: Own computation

According to the above table all of the respondents who hold diploma and above were participated in controlling the impact of population growth on forest. About 80 percent of farmers who attained grade 11 and 12 were engaged in controlling the impact of population growth on forest development.

As per the survey data 85 of the respondents s were respond as the district administration have taken corrective measures for cutting trees either by planting other trees or legal action. While 4 of the respondents reflect their idea as there is no action taken by the district for cutting trees.

Table 4.3. Purpose for using forest

		Purpose for using forest				Total
		Household consumption	Manufacturing materials	Energy such as charcoal	Agricultural materials	
Sex of respondents	Female	1	6	1	1	9
	Male	21	28	11	20	80
Total		22	34	12	21	89

Source: Own computation

According to the above table 34 of respondents' foremost use the forest for manufacturing materials while 22 of them foremost use forest to run their household consumption. By the same way 12 of the respondents use the forest for fulfilling their energy sources such as charcoal and 21 of them mainly use forest for attaining agricultural materials.

As per the survey result shows 84 of the respondents use family planning services in reducing the impact of population growth on the forestry development. While 5 of them have no idea of controlling population growth as it has no impact on forest management.

Table 4.4. Perception of environmental problem

		Perception of environmental problem			Total
		Increasing	Decreasing	No change	
Sex of respondents	Female	8	1	0	9
	Male	73	5	2	80
Total		81	6	2	89

Source: Own computation

From the above table we can see that 81 of the respondents perceive increment in environmental problem, 6 of them perceive decrement in environmental problem and 2 of them perceive as there is no change in environmental problem.

Graph 4.1 Farmers perception of environmental problem

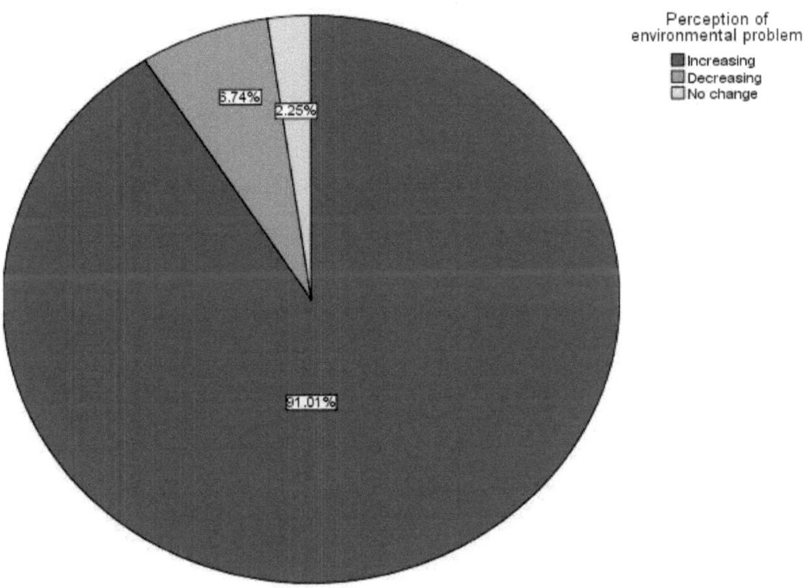

4.2. Causes and Consequences of Deforestation

According to Johnson and Chenje (2008), deforestation is when a once forested area gets removed for the use of agriculture, pasture, urban development, logging, or wasteland. When this destruction of a natural environment occurs, a degraded ecosystem is likely to follow. A degraded ecosystem is the result of a loss of habitat and a reduction in biodiversity. Also, the erosion of soil is usually a consequence of deforestation.

People have been deforesting the Earth for thousands of years, primarily to clear land for crops or livestock (Julie M, 2009). Although tropical forests are largely confined to developing countries, they aren't just meeting local or national needs; economic globalization means that the needs and wants of the global population are bearing down on them as well. According to Julie M (2009) the direct causes of deforestation are agricultural expansion, wood extraction (e.g., logging or wood harvest for domestic fuel or charcoal), and infrastructure expansion such as road building and urbanization. Rarely is there a single direct cause for deforestation. Most often, multiple processes work simultaneously or sequentially to cause deforestation.

Deforestation releases carbon, principally as CO_2, to the atmosphere as the organic carbon stored in trees and soil is oxidized through burning and decay. Other greenhouse gases, such as CH_4 and N_2O, are also emitted as a result of the conversion of forests to agricultural lands (R. A. Houghton, 2004).

Table 4.5 Major cause of deforestation

		Major cause of deforestation				Total
		Overgrazing	Cutting down trees	Soil degradation	Over cultivation	
Sex of respondents	Female	1	3	3	2	9
	Male	4	48	19	9	80
Total		5	51	22	11	89

Source: Own computation

Graph 4.2 Major cause of deforestation

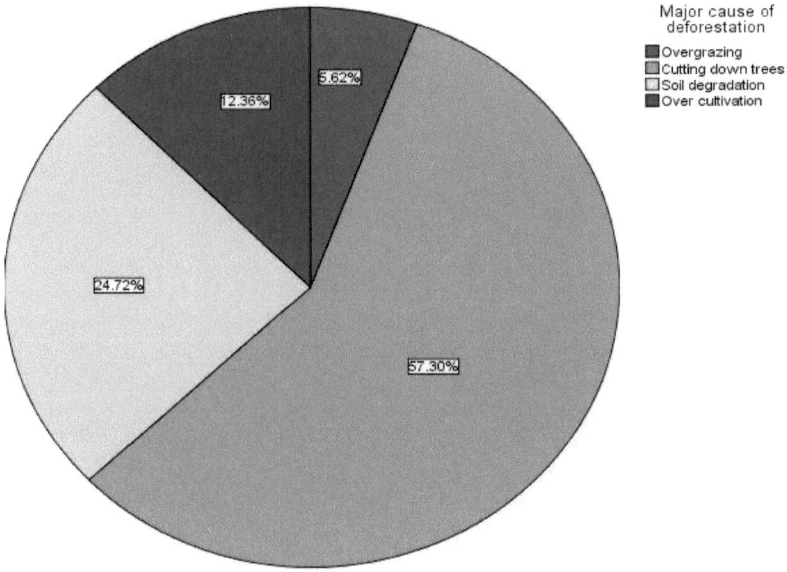

As per the result of the survey of this study the major consequences of deforestation in the study area are desertification and low agricultural productivity.

Table 4.6 Major consequence of deforestation

		Major consequence of deforestation		Total
		Desertification	Low agricultural production	
Sex of respondents	Female	6	3	9
	Male	72	8	80
Total		78	11	89

Source: Own computation

Graph 4.3 Major consequences of deforestation

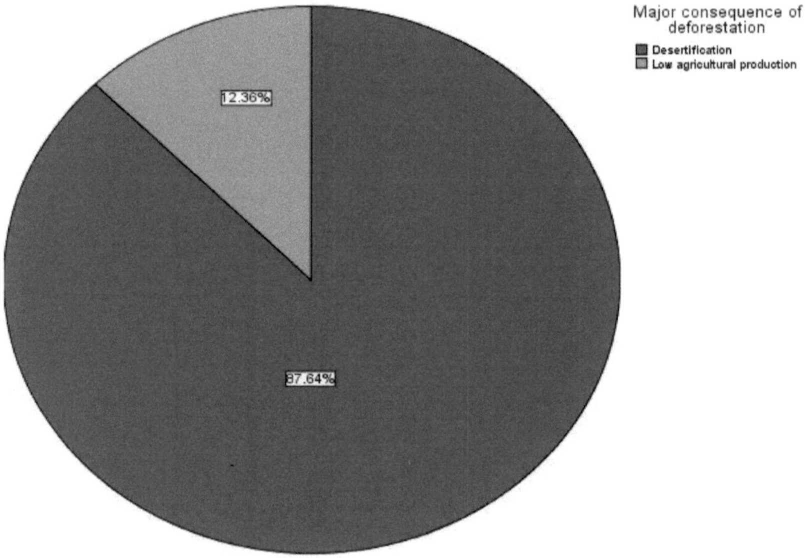

All of the respondents were well aware of the impact of the population growth on deforestation in the study area. Some of the respondents were also mentioned the real situation exist before and after the settlement of the settlers since the number of the population was increased after their settlement.

About 88 of the respondents were used forest in order to satisfy their energy requirement as a wood. This is the major problem that contributes to increment in deforestation. When I was conducted interview with respondents the local and district administration was not engaged in taking measures on this area.

Table 4.7 Common energy usage

		Common energy usage		Total
		Wood	Charcoal	
Sex of respondents	Female	9	0	9
	Male	79	1	80
Total		88	1	89

Source: Own computation

As explained on the graph below, the major causes of deforestation in Haro Limmu district was cutting trees, over cultivation, soil degradation and overgrazing. Overgrazing is more common in drier areas of the tropics where pastures degraded by overgrazing are subject to soil erosion.

CHAPTER 5

5. Summary, Conclusions and Recommendations

5.1. Summary

Deforestation is a growing problem in many parts of the tropical world and one of the affected countries is Ethiopia. The rapid deforestation or the declining of forest accompanied by variety of other environmental problems is caused by agricultural expansion and wood gathering. The major factors contributing to deforestation are poverty, low level of income and population growth.

The general objective of this study is to assess the effect of population growth on forest resource in East Wollega Zone in general and Haro Limu woreda in particular. The data used for the study were collected from 89 farm households heads drawn from the four kebeles of Haro Limmu district. Probability proportional to size sampling technique was employed to select the farm households from four peasant associations, which were selected by random sampling techniques. Primary data were collected using a structured questionnaire. In addition, secondary data were extracted from relevant sources to supplement the data obtained from the survey.

In carrying out this study, it can be hypothesized that most of the population growth members participating in forestry which may be affect forest or not and there is impact of population growth on forestry in this area to decided the effect of the forest beneficiary.

5.2. Conclusions

Although Ethiopia has a large population that is increasing rapidly, the constant chopping down of forested areas to convert to agricultural land needs to either decline or use more sustainable methods. Converting forests to farms is obviously necessary for food production, but soon the environmental impacts resulting from the degradation of land will be too much to bear for the people in Haro Limu and other neighboring districts. The impact of demographic changes on forests and the environment is often discussed in terms of biological carrying capacity, i.e. the maximum number of individuals that a resource can sustain. The rapid deforestation or the declining of forest accompanied by variety of other environmental problems is caused by agricultural expansion and wood gathering. The major factors contributing to deforestation are poverty, low level of income and population growth.

Population growth has contribution immensely to resources degradation especially as it affect the forest ecosystem. It was observed that forest resources in Haro Limmu district are under a serious depletion due to population growth and human encroachment in the forest ecosystem. However, there is urgent need for the various stakeholders in environmental resource management to provide a mechanism that would prevent the forest ecosystem from further depletion in the area.

As per the result of survey most of the respondents have been participating on forestry development. This is towards the hypothesized concept which explains most of the population growth members participating in forestry which may be affect forest or not. It is also known that the population growth huge impact on forestry development in the ways of expanding agricultural land, using wood as energy sources and satisfying the input requirements in agricultural activity.

The other result of the study is the linkage between population growth and deforestation. In this case the respondents of the study responds that people was engaged on cutting down trees and the soil was degraded as a result of overgrazing and over cultivation. Therefore there is deforestation as a result of continuous increment of population.

5.3. Recommendations

The rate of human disturbances on the forest ecosystem in Haro Limmu district has generated serious environmental problems in the area. Therefore, to avert this problem, the following measures are hereby recommended:-

- Population increase should be checked through family planning methods, this would cut down the large family size.
- Proper environmental education should be imparted to the people so that the people can appreciate the importance of sustainable environment.
- The inhabitants should be encouraged to carry out wildlife and ecosystem conservation.
- Since almost all of the respondents were using wood as energy source, this has great impact on forest development. So, the local and district administration should have take a measures on this activity as they are taking measures on those who have cut the forest for expansion of agricultural land.

6. REFERENCES

Amare S. M., (2013) Population and environment interaction: the case of gilgel abbay catchment, northwestern Ethiopia, *Journal of Environmental Research and Management* Vol. 4(1). pp. 0153-0162, January, 2013

Angelsen, A.; Shitindi, E. F. K. and Aaarrestad, J. (1999) Why do farmers expand their land into forests? Theories and evidence from Tanzania. *Environment and Development Economics* 4: 313-31.

Barraclough, S. and Ghimire, K. B. (2000) *Agricultural Expansion and Tropical Deforestation*, Earthscan

Bijendra Basnyat (2009) Impacts of Demographic Changes on Forests and Forestry in Asia and the Pacific, Food and Agriculture Organization of the United Nations Regional Office for Asia and the Pacific, Bangkok

Carla B., Gezu B. , Amare I., and Arianna O., (2001) Population Growth and Environment in Ethiopia, Addis Ababa, Ethiopia

Chakravarty, S. K. Ghosh, C. P. Suresh, A. N. Dey and Gopal Shukla (2012). Deforestation: Causes, Effects and Control Strategies, Global Perspectives on Sustainable Forest Management, Dr. Clement A. Okia (Ed.), ISBN: 978-953-51-0569-5, InTech

Dealon R.T (1993) Assessing the Relationship between Government Policy and Deforestation, *Journal of Environmental, Economic and Management*

Encarta (2009) Microsoft Encarta, Microsoft Corporation

FAO (2001) Soil fertility management in support of food security in sub-Saharan Africa, Food and Agriculture Organization of the United Nations, Rome, 2001

Geoffrey D. Dabelko (2011) Population and Environment Connections, Council on Foreign Relations (CFR) *Women and Foreign Policy program, United States of America*

HLDFEDO (2011) Physical and Socio-Economic Profile of Haro Limu District

Hamdallah Z., (2001) The Value of Forest Ecosystems, Secretariat of the Convention on Biological Diversity ISBN 90-907-2111-9

Johnson, Rebecca L. and Munyaradzi Chenje (2008) *Africa: Atlas of our Changing Environment*

(Nairobi, Kenya: United Nations Environment Programme (UNEP)); available at: http://www.unep.org/dewa/africa/AfricaAtlas/PDF/en/Africa_Atlas_Full_en.pdf.

Julie Mattern (2009).*Causes of Deforestation: Direct Causes*

Myers, N. (1994) Tropical deforestation: rates and patterns. In: *The Causes of Tropical of Tropical Deforestation. The economic and statistical analysis of factors giving rise to the loss of the tropical forest*, eds. Brown, K. and Pearce, D. pp 27-40. UCL Press

Negash Mamo (2003) Journal of Natural Resource Volume 2, 273

R. A. Houghton (2004). Tropical deforestation as a source of greenhouse gas emissions, The Woods Hole Research Center, Woods Hole, Massachusetts, USA

Steve J., Ruth N., Neil J. and Tom E., (2003) The High Conservation Value Forest Toolkit

Van Kooten, G. C. and Bulte, E. H.(2000) *The economics of nature: managing biological assets*, Blackwells

APPENDIX

I. Appendix one Questionnaire (English)

The objective of this questionnaire is to collect information related to effect of population growth on forest resource in Haro Limu Districts. The study will be conducted only for academic purpose. Hence, I request your honest and fair responses to fill up this questionnaire.

Questionnaire

1) Sex of the household head

   0 ☐ Female 1 ☐ Male

2) What is your marital status?

   0 ☐ Never married 1 ☐ Married 2 ☐ Divorced 3 ☐ Separated

3) What is your religion?

   0 ☐ Orthodox 1 ☐ Protestant 2 ☐ Muslim 3 ☐ Catholic

4) Level of education of the household head

   0 ☐ Illiterate 1 ☐ Less than or equal to grade four 2 ☐ Grade 5 - 8

   3 ☐ Grade 9 -10 4 ☐ Grade 11 – 12 5 ☐ Certificate 6 ☐ Diploma and above

5) How many years of experience do you have in farming? ----------------

6) Household size ----------------

7) Age of the Respondent Household ______________

8) Have you controlled the impact of population growth on forest in your woreda?

   0 ☐ Yes 1 ☐ No

9) If your answer for question number 8 is yes how?

   a) ____

   b) ____

   c) ____

   d) ____

10) For what purpose do you use forest?

   0 ☐ Household consumption 1 ☐ Manufacturing materials

   2 ☐ Energy such as charcoal 3 ☐ Agricultural materials

4 ☐ Others (specify) ________________________

11) Do you conserve forest from damage?

0 ☐ Yes 1 ☐ No

12) If your answer for question number 11 is yes how?

a) ____

b) ____

c) ____

d) ____

13) Does your woreda or peasant association make to plan other tree when they cut the other?

0 ☐ Yes 1 ☐ No

14) Do you use family planning services as a remedy to reduce population growth?

0 ☐ Yes 1 ☐ No

15) How does the household perceive the environmental problem since starting farming as compared to the past?

0 ☐ Increasing 1 ☐ Decreasing 2 ☐ No change 3 ☐ Do not know

16) From the following which one is the major cause of deforestation in your area?

0 ☐ Overgrazing 1 ☐ Cutting down trees 2 ☐ Soil degradation

3 ☐ Over cultivation 4 ☐ other (specify) ______________

17) From the following which one is the major consequence of deforestation in your area?

0 ☐ Poverty 1 ☐ Disease 2 ☐ Desertification

3 ☐ Low agricultural production 4 ☐ other (specify) ______________

18) Do you think that population growth may have impact of deforestation?

0 ☐ Yes 1 ☐ No

19) How do you observe population growth in your peasant association?

0 ☐ Increasing 1 ☐ Decreasing 2 ☐ No change 3 ☐ Do not know

20) What are the most commonly used types of energy for your household?

0 ☐ Wood ☐ 1 Shrubs and leaves ☐ 2 Crop residue ☐ 3 ☐ Dung 4 ☐ Charcoal 5 Other (Specify) --------

21) What are the measures taken in reducing deforestation in your area?

a) ______________________________

b) ______________________________

c) ______________________________

d) ______________________________

e) ______________________________

f) ______________________________

22) Why people cut down trees and or forest?

a) ______________________________

b) ______________________________

c) ______________________________

d) ______________________________

e) ______________________________

f) ______________________________

II. Appendix Two Questionnaire (Afan Oromo)

Kaayyoo ani gaaffilee gadii kana itti isin gaafadhu yaada dhiibbaa guddinni uummaataa kununsa bosonaa irratti qabu ilaaluuf yoo ta'u, kayyoon qorrannoo koos aanaa Haroo Limmuu keessatti dhiibbaa baayyinni uummataa bosona irratti qabu duwwaa ilaaluudha. Qorannaan kunis kayyoo galma ga'insa barnoota koo duwwaatiif akka oolu isin hubachiisaa, yaada keessan bilisaan akka anaaf laattan isin gaafadha.

Gaaffilee

1) Saala abbaa warraa

   0 ☐ Dhalaa   1 ☐ Dhiira

2) Haala gaa'ela abbaa warraa?

   1 ☐ Kan hin fuune   1 ☐ Kan fuudhe   2 ☐ Kan irraa du'e   3 ☐ Kan addaan ba'an

3) Amantii abbaa warraa?

   1 ☐ Ortodoksii   1 ☐ Pirotestaantii   2 ☐ Muusilima   3 ☐ Kaatolikii

4) Sadarkaa barnoota abbaa warraa

   1 ☐ Hin baranne   1 ☐ Kutaa afurii fi isaa gadi   2 ☐ Kutaa 5 - 8

   3 ☐ Kutaa 9 -10   4 ☐ Kutaa 11 – 12   5 ☐ Setifikeetii   6 ☐ Diploomaa ol

5) Qonnaa irratti erga bobbaatanii waggaa meeqa? ----------------

6) Baayyinni maatii keessanii meeqa ----------------

7) Umuriin keessan meeqa? ____________

8) Dhiibbaa baayyinni uummattaa bosona irratti qabu hambisuu keessatti hirmaannaa qabduu?

   0 ☐ Eeyyee   1 ☐ Lakkii

9) Yoo deebiin keessan gaaffii 8ffaa Eeyyee ta'e akkamiin?

   a) ____
   b) ____
   c) ____
   d) ____

10) Bosona caalmaatti maaliif fayyadamtuu?

   1 ☐ Nyaata maatiif   1 ☐ Meeshaalee ijaarsaatiif

   2 ☐ Qoraanii fi qasalaaf   3 ☐ Meeshaalee qonnaatiif

4 ☐ kan biroo (ibsi) ______________________________

11) Aanaan ykn gandi keessan kun namni muka muree argame akka dhaabuuf ni dirqamsiisuu?

0 ☐ Eeyyee 1 ☐ Lakkii

12) Baayyina ummataa xiqqeessuuf karoora maatii akka meeshaatti ni fayyadamtuu?

0 ☐ Eeyyee 1 ☐ Lakkii

13) Erga qonnaa eegaltee rakkoo jijjiirama haala qilleensaa kana akkamiin hubataa jirta?

1 ☐ Dabalaa jira 1 ☐ Hir'ataa jira 2 ☐ Hin jijjiiramu 3 ☐ Hin beeku

14) Kan armaan gadii keessaa sababni gammoojummaan naannoo keessanitti akka mul'atuuf godhaa jiru inni guddaan kamii?

1 ☐ Dheechisa 1 ☐ Bosona muruu 2 ☐ Dhiqama biyyoo

3 ☐ Deeddebianii oomishuu 4 ☐ Kan biroo (ibsi) ________________

15) Kan armaan gadii keessaa dhiibbaa gammoojummaan naannoo keessanitti akka mul'atuuf godhaa jiru inni guddaan kamii?

0 ☐ Hiyyummaa 1 ☐ Dhukkuba 2 ☐ ho'ina qilleensaa

3 ☐ Oomishini xiqqaachuu 4 ☐ kan biroo (ibsi) ________________

16) Baayyinni uummataa gammoojummaa irratti dhiibbaa qaba jettee ni yaaddaa?

0 ☐ Eeyyee 1 ☐ Lakkii

17) Naannoo keetti baayyina uummataa akkamiin hubattaa?

0 ☐ Dabalaa jira 1 ☐ Hir'ataa jira 2 ☐ Hin jijjiiramu 3 ☐ Hin beeku

18) Mana keessatti human annisa kamiin caalmaatti fayyadamtuu?

0 ☐ Qoraan 1 ☐ Baalaa fi offaa 2 ☐ Offaa dhangaa

3 ☐ Dikee loonii 4 ☐ Charcoal 5 ☐ Kan biroo (ibsi) ------------

19) Naanno keetti gammoojummaa kana xiqeessuuf mallii ati fudhattee maalii?

a) ______________________________
b) ______________________________
c) ______________________________
d) ______________________________
e) ______________________________
f) ______________________________

20) Naannoo keetti namni maaliif bosona ciraa?

1 ______________________________
2 ______________________________
3 ______________________________

4 ____
____